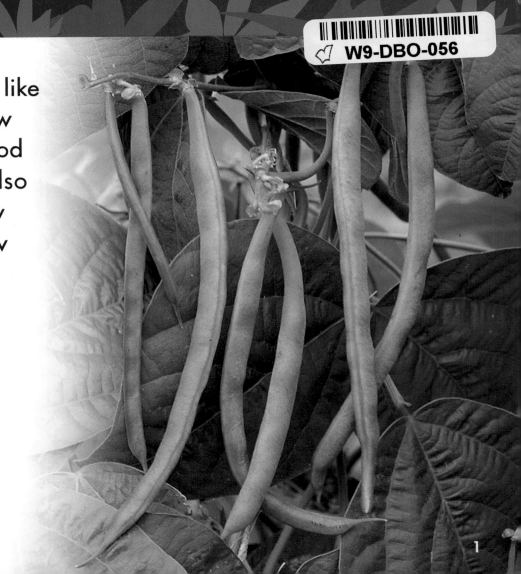

What would the garden be like without beans? People grow beans because they are good to eat. But the bean pods also have another purpose. They contain seeds that will grow into new bean plants.

Pods are the fruit of the bean plants. If you run your fingers up and down a bean pod, you will feel bumps. Each bump is a seed. Split open a pod to see its seeds.

If you put a bean seed in soil and give it water, something magical will happen.

Inside each seed is a tiny plant, or embryo. The seed also has two large white sections called seed leaves. They are full of food for the embryo.

In moist soil, the seed slowly swells with water. The embryo's tiny root begins to grow. The root breaks through the outer shell, or seed coat, and pushes downward into the soil. The seed has germinated.

A pale green shoot begins to grow. It pushes up and forces the seed leaves and seed coat apart. The seed coat falls to the ground.

The shoot has
two new leaves.
They spread open
to face the sun.

The seedling keeps growing. More leaves form. The leaves make food for the whole plant. To make food, they use water from the ground, light from the sun, and carbon dioxide from the air.

Beetles and other insect pests come to eat the bean leaves. If the insects do not eat too many leaves, the plant will be strong enough to keep growing. When it is big enough, flower buds form. Soon they open into flowers.

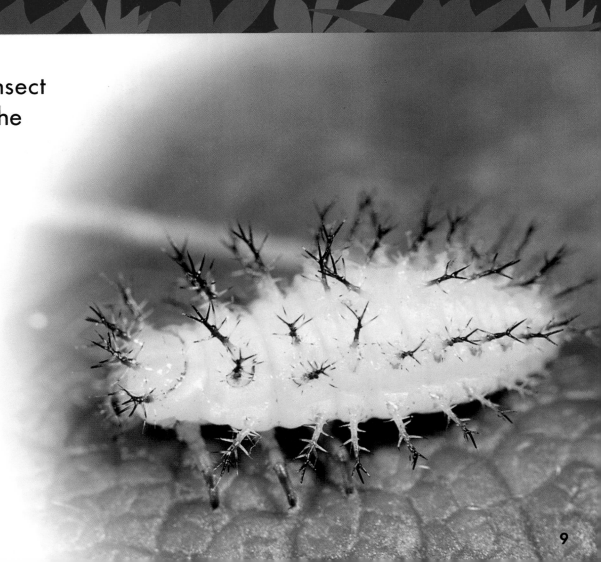

Bees and other insects visit the flowers to sip the sweet nectar. A yellow dust called pollen sticks to the insects' bodies. When the insects move to other flowers, the pollen rubs off. It fertilizes tiny eggs deep inside the flower.

The eggs begin to grow into new seeds, and a bean pod forms around them.

The flowers wither and drop off, but the pod keeps growing. The seeds inside get bigger. Fruit and seeds are good to eat, so people pick the pods. If they save some seeds and plant them next spring, the seeds will grow into new bean plants.

Can you put these in order?